The book of being Chimp

The Book of Being Chimp
Published in Great Britain in 2024 by Graffeg Limited.

Written by Adrian Cale copyright © 2024.
Photography by Adrian Cale copyright © 2024.
Produced by Graffeg Limited copyright © 2024.

Graffeg Limited, 15 Neptune Court, Cardiff, CF24 5PJ, Wales, UK. www.graffeg.com.

Adrian Cale is hereby identified as the author of this work in accordance with section 77 of the Copyright, Designs and Patents Act 1988.

A CIP Catalogue record for this book is available from the British Library.

All rights reserved. No part of this publication may be reproduced, stored in a retrieval system or transmitted, in any form or by any means, electronic, mechanical, photocopying, recording or otherwise, without the prior permission of the publishers.

ISBN 9781802587876

1 2 3 4 5 6 7 8 9

The book of being Chimp

Adrian Cale

Foreword by Jane Goodall

GRAFFEG

Contents

6	Foreword
9	Introduction
11	**Part 1 – Being a chimp**
12	Being a chimpanzee
14	Being like us
17	Being at home
18	Being different
20	Being a baby
22	Being a mum
24	Being a family
26	Being a community
28	Being the boss
30	Being social
32	Being groomed
35	Being a toddler
36	Being funny
38	Being able to talk to each other
40	Being clever
42	Being chilled
44	Being almost grown up
47	Being handy
48	Being able to walk on your knuckles

50	Being able to climb
52	Being hungry
54	Being really grown up
56	Being taken
58	Being abandoned
61	**Part 2 – Being helped to be a chimp**
62	Being rescued
64	Being seen by the vet
67	Being sad
68	Being supported
71	Being bottle-fed
73	Being cleaned
74	Being dependent
77	Being scared
78	Being encouraged
81	Being sleepy
82	Being happy
85	Being ticklish
86	Being hugged
88	Being curious
91	Being independent
92	Being endangered
96	About the author

Foreword

The Book of Being Chimp is Adrian Cale's hugely entertaining celebration of chimpanzees – our closest, brightest, most sensitive, family-loving and loyal animal relatives, with some of the most adorable infants in the animal kingdom.

In this book, he brings together some truly beautiful pictures with writing that is packed with facts in a light and humorous way. If chimps themselves could read this might well be their own informative how-to-grow-up guide. These delightful pictures and simple text tell you so much about chimps and why we should protect them.

Through endearing photos and informative and often witty prose, this book tells the story of chimpanzee life – from an early age, learning who's who in the family, how to behave and make friends, building relationships that might last a lifetime, learning what fruits and leaves are good to eat and how to make and use tools for feeding.

We go from finding out how chimps grow up to exploring what it takes for them to be a grown-up and the new challenges that come with that, not least the effect that people are having on them. Sanctuaries take in more and more orphan chimpanzees and this book also celebrates the dedicated work they do in saving them.

A family book for all ages, *The Book of Being Chimp* is emotive, informative, inspirational and enchanting, with stunning photographs and clear writing throughout. It will not only help inspire the young chimp conservationists of the future but will also encourage those already initiated in the wonderful world of being a chimpanzee.

Jane Goodall PhD, DBE
Founder, the Jane Goodall Institute
& UN Messenger of Peace

Introduction

Chimps and the lives they lead are as complex as they are chaotic. Through my pictures and prose, this book aims to unravel the complex and calm the chaotic.

World-renowned chimp expert Dr Jane Goodall described this as a chimp's very own how-to-grow-up guide. Alas, we'll never know. Chimps can't read, but then again, I can't walk on my knuckles, so we'll call it even.

Some of it might seem a little hard-hitting, because it has to be, but so, so much more isn't. They are one of our closest living animal relatives – and for that fact alone I think they deserve to be celebrated.

This book does just that. It is my personal homage to the chimps that fascinated me as a youngster and the chimps that continue to fascinate me today.

Hopefully, by the end of it, you'll be brimming with knowledge and rooting for chimps everywhere.

Adrian Cale

Part 1
Being a chimp

Being a chimpanzee

Chimp is less of a mouthful to say than chimpanzee. Being a chimp is not the same as being a monkey. They are often called monkeys by mistake but, although related, chimps are definitely not monkeys.

For a start, monkeys have a tail and chimps do not, and monkeys come in many different shapes and sizes. They can be as small as a rat or as big as a big dog, have long hair, short hair or multi-coloured hair, and live on several different continents in the world. But chimps do not.

Chimps only live in Africa and have long arms that are as strong as anything and perfect for climbing trees. They have hands that are almost the same as ours and they are very clever. So clever, in fact, that they are one of the most intelligent animals on the planet. If there were such a thing as animal school, chimpanzees would be top of the class. From now on, no more monkey-malarkey – they are chimps, wonderful, glorious chimps in all their chimpy brilliance.

Chimps can recognise themselves in the mirror.

Being like us

Chimps are members of the family. Every living thing is made up of DNA, the stuff that makes all the bits and bobs work and look like they do. Different animals have different DNA, but chimps are special. Around 98 per cent of their DNA is the same as ours, which makes them one of our closest living animal relatives.

Along with gorillas, orangutans and bonobos, chimps are great apes, an exclusive club where everybody has very similar DNA, with chimps and bonobos just that bit closer to us. Great apes sit at the top of the primate tree, one big animal family that also includes gibbons (lesser apes), monkeys and lemurs.

Chimps like each other's company and live together in groups and communities. They are great mums and care for their youngsters for a very long time. Loving families look out for each other like we do and, being self-aware and emotional, things matter to chimps. They like to play, laugh, cuddle, hold hands, make friends, fall out and fight. Chimps are indeed just like us.

Being so closely related, you might think we look nothing like them. But look again. Those eyes, the hands, the feet, the arms, those ears and that smile. Suddenly, we can see a lot of ourselves in them – apart from the fact they are much, much hairier.

Being at home

Chimps live in forests throughout West and Central Africa and some parts of East Africa. Depending on their location, how they live and in what habitat can vary greatly.

A chimp's preferred home is lush rainforest, where they are found in the greatest numbers, but some must settle for less luxurious accommodation. Other habitats include savannahs where trees are more spread out, secondary forests where trees have re-grown, and arid areas where trees are more a bonus. You also find chimps living in woodland, bamboo, and swamp forests. When it comes to making themselves at home, chimps are admirably adaptable.

Thumbs up for forest life!

Being different

All chimps look different to each other. In some cases, very different. One might have a big nose and another exceptionally big ears. They can have freckles, pale skin, dark skin, or a blotchy mix of both. Some are big while others stay lean and lanky no matter how much they eat. A typical chimp group is a varied cast of instantly recognisable characters.

Most have black hair while a few are a striking chocolate brown. Some are a blend of black and brown and anything in between, and others are dark grey, verging on black.

Many start to go grey anyway, particularly around the mouth and worn as a beard, or on the back as a salt and pepper look of distinction.

There are also those that struggle to hang on to a full head of hair at all. As they get older, chimps can go bald, and both males and females are just as likely to become follicly challenged.

Being a baby

Baby chimps become celebrities almost as soon as they are born, an overnight sensation everybody wants to meet. Friends and family pop by to say hello and sneak a cuddle if they are lucky, while mum does her best to keep the adoring fans happy and over-boisterous ones at arm's length. In chimp society, babies, also known as infants, are very special indeed. They are the glue that holds the group together.

New babies are very easy to spot. They look like pocket-sized versions of their future self, with skinny arms in need of growing into and big eyes with a bewildered look that could melt the coldest of hearts.

Their faces, hands, ears and feet – all the fleshy bits – are very pale compared to adults, and they have a tuft of white hair on their backside that really stands out. This tuft tells all the chimps in the group that they are a baby and might not always know what they're doing, so please go easy.

The tuft disappears as they get older, at which point young chimps have no excuse for bad behaviour and are more prone to a clip around the ear when others run out of patience.

Being a mum

It's easy to tell when a chimp is ready to become a mum – she grows a rather large pink bottom. This tends to start from around ten years old and makes her very popular with the boys. Chimps give birth after about an eight-month pregnancy and won't have another baby for perhaps the next four or five years.

Being a chimp mum is a full-time job. For the first year, she keeps her baby close to her day and night, even sleeping with at least one hand in constant contact. She will teach essential life skills and provide transport in the form of a primate piggyback. It's a bit like riding a horse without a saddle. Luckily, babies have a very strong grip and no problem clinging to mum's hair as they ride her around the forest.

Chimp mums don't have many babies. A super-smart animal needs time to develop properly and chimp mums invest as much time as necessary to ensure that happens. So even though chimps can live for over 50 years, mums usually only have five or six babies in their lifetime.

Mum and baby have a very special bond that will last a lifetime.

Being a family

Family means so much to chimps. They live in extended family groups where older sisters are known to adopt their younger siblings if something happens to mum. Aunties might also step in and adopt when necessary.

Youngsters learn so much from their big brothers and sisters, spending time with them and watching and practising what they do. Grandma could also be around to share her wisdom or a shoulder to cry on. Sibling rivalry is a thing too, with jealousy and attention-seeking a reminder of just how closely related we are.

Brothers don't always stick to the family code. As they grow up, they might fall out over who becomes the leader of a group. But when push comes to shove, the family are there for each other. Whether it's standing up to others, boosting confidence, cleaning wounds or sharing food, chimp families are a rollercoaster ride of love, loyalty, fighting, frolicking, sharing and caring.

Large families include multi-generations and the bigger the family the more influence it has in their community. Some chimps do leave and move on to seek their fortune elsewhere. Years might pass before they meet up with family members again, but when they do it's often as if they've never been away.

Being a community

Chimps live in communities of around 15 to over 100 individuals. They are led by the highest-ranking chimp, the alpha male, who keeps control of them all. These communities pass on years of know-how from one generation to the next. New ways of doing things often pop up along the way and are subsequently copied by the rest of the community until they become the norm, the group culture. That is why you see chimps in different parts of Africa doing different things; a chimp community in Uganda may well develop a unique set of skills to those in Tanzania, and vice versa.

Chimp communities are seldom all together in the same place at the same time, preferring instead to hang out in smaller groups, or subgroups, of a select few. It's better to spend the day in the company of a good friend than make small talk with a boring chimp whose personality is as wooden as the trees they climb.

These subgroups are like neighbours living alongside one another in a large town, frequently bumping into each other down the shops and exchanging niceties before heading home to their nearest and dearest. Sometimes subgroups will spend quality time together, and at other times they prefer their own company. It's known as a fission-fusion society. These groups are flexible too, with members coming and going over time.

Being the boss

There can only be one leader of a chimp community – the big boss, the main man – the alpha male.

Bigger and stronger than females, males can weigh around 80 to 130 pounds. That's a pretty hefty weight packed into a four to five-feet-high frame when standing up. Being much stronger than the strongest man, they are a force to be reckoned with.

The boss has many perks. He tends to have more babies than other males, the best seat at the dinner table and the first choice of food. The best bosses are treated like rock stars, so it's no surprise other males want what he has.

Keeping control within the community, head-turning displays, consideration, grooming, supporting allies and fighting rivals are all part of the job description. He also needs to show a calm and tender side with the females. A constant brawn before brains approach can leave the ladies cold, so when another chimp charmer looking to take over catches their eye it doesn't take much for them to ditch a bully boss.

Chimp leadership fights can sometimes be brutal. After years of climbing the chimp social ladder, the way down for a deposed boss can be quicker than a fireman's pole.

Being social

Chimps like to have friends. Best friends help each other grow up, and really best friends often support each other for the rest of their lives.

They have a very complex social life that includes jealousy and rivalry, so having many friends that have your back is advisable because politics rules chimp life. In chimp society, there are the haves and the have-nots. The ones that have are political movers and shakers, usually bright, often crafty, and good at climbing the social ladder by themselves or by supporting others who are better at it than they are. The have-nots are consigned to looking over their shoulder, waiting their turn or being whacked on a whim to remind them of their place. But the hierarchy is not set in stone, with political power plays never far from a chimp on the lookout for a lifestyle change.

Being an adult with plenty of pals is about as good as it gets. The most popular ones might become future leaders of a community and a strong network of chimp chums helps them get there and stay there. Some can be quite sneaky and change friends as often as we change our socks, but it's the long-term ones that really matter, the ones that are there through thick and thin, happy to share lunch with a friend or stand shoulder to shoulder against a bully.

Having a wheelbarrow race is a perfectly sociable thing for young chimps to do.

Being groomed

Chimps have a 'you scratch my back and I'll scratch yours' approach to life. Part of being super-social means spending an awful lot of time taking care of each other's hair, or rather what's living in or stuck in each other's hair.

Grooming is a serious business, with the hairdresser concentrating so intently you'd think there was an exam at the end of it. They groom pretty much every part of the body, though not all in one session, and use their fingers and lips as a kind of brush and comb. Chimps are especially diligent when it comes to grooming, skilfully parting individual hairs and extracting all kinds of muck and more. Dirt, dead skin and parasites are all fair game for a nimble-fingered groomer.

Grooming is not just about keeping clean though. It is an essential part of bonding, helping to form those all-important alliances and friendships that are key to chimp life. It's like an unspoken code that enhances existing relationships, forges new ones, and comforts nervous or wound-up chimps having a bad day. It's also a perfect way to stay in the boss's good books, who will in turn groom others as a bit of give and take to keep them on his side.

Chimps generally offer one-on-one grooming sessions but will take group bookings.

Being a toddler

From the age of about three, chimp toddlers start to venture out on their own, albeit for short distances to begin with. There's a big wide world out there to explore, but it's not always safe, so mum keeps an ever-watchful eye on them and will step in if things get too heavy.

Of course, three-year-old chimps are not really called toddlers – that's a human thing – they are still called infants, but you'd be forgiven for seeing the similarities to us as toddlers in much of what they do. They are inquisitive, full of energy and into everything, especially playing.

Playtime can be anything from rough and tumbles with each other, sneaking off to explore behind mum's back, winding up some of the older chimps for fun, or working out the best launch pad to jump on a sleeping grandad.

As infants become more independent, there's also no need for them to ride on mum's back all the time, unless, of course, they want to rest their little legs.

Being funny

Chimps have a natural ability to make us laugh and can pull the kinds of faces a mime artist would be proud of. This is more by luck than judgement because we see so much of our own behaviour in them. To us, some appear to have funnier bones than others.

Take the baggy lip routine, for example. For this, you have to be born with a putty pout to pull it off. It's a rather impressive combination of looking lost and looking amazed in equal measure.

Just poking your tongue out to anyone watching is a cheeky but highly effective way to raise a smile.

Then there's the folding of the bottom lip inside out and holding the pose just long enough to get a reaction but not long enough to dry your gums out.

Having a flexible face is actually very important for chimps. It allows them to communicate with each other using facial expressions, and so what seems like funny faces to us are simply chimp chit-chat in most cases.

Being able to talk to each other

Chimps 'talk' to each other, using many ways to do so. From simple gestures and subtle looks to full-on displays and acts of affection, chimp chat involves a bit of everything. With an impressive vocal range and the ability to pull a face for all occasions, chimps are constantly communicating.

Facial expressions

Different grins say they are scared or excited. They sneer, pout and grimace, and males threaten each other by pulling their lips in like an old man mouthing without his false teeth. Chimps have a play face too, showing the bottom teeth while keeping the top ones hidden. They like to smile and laugh, often rolling around in hysterics at what looks like the funniest chimp joke ever told.

Physical

Chimps cuddle to show affection and give reassurance, and kiss and hold hands in friendship. A mum will stick out a leg and show the sole of her foot to her baby as a way of saying 'jump on my back', and raising an arm in the air or staring another chimp out means things are about to kick off. Males will display by throwing their weight around, swaying and swaggering, shaking branches and stamping their feet. And always with every hair on their body standing up as if they've used far too much hair gel.

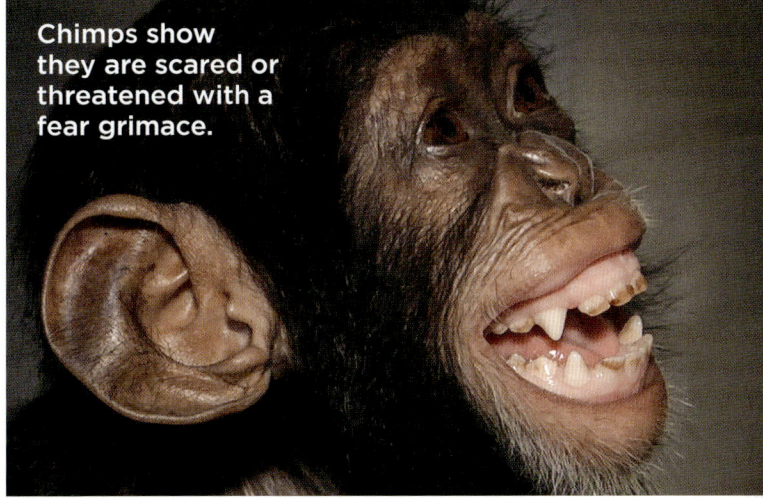

Chimps show they are scared or threatened with a fear grimace.

Vocal sounds

Pant-hoots are a chimp's own unique voice; no two are the same so they know exactly who's hooting who. Chimp chatterboxes also whimper, squeak, grunt, bark and scream very loudly. All ways of telling each other how they feel, what they need or what is happening at any time.

Being clever

Chimps are brilliant students. It helps that they have large brains and a knack for problem-solving and remembering things, like the best trees for a fruit feast at certain times of the year. They have even learnt how to make tools. People used to think it was only us who made tools to get a job done, but it turns out that chimps are pretty good at it too and can make and use many different types. However, while most take to tool use like a duck to water, others are as sharp as a spoon, never quite getting it no matter how many times they try.

Basic tools include chewing leaves and making a sponge to get water from hard-to-reach places, or simply holding leaves over their head like an umbrella. Two of their most successful inventions are making fishing gear to catch insects and turning rocks into useful hammers.

Fishing rod

Skilled chimps find a stick that must be just the right width and pull off the leaves and other sticky-out bits to fashion a perfectly shaped fishing rod. Ideal for dipping into termite nests to catch a creepy-crawly snack.

Cracked it!

Hammer and anvil

A tool made up of two rocks – one perfectly shaped to sit a nut on, the other perfectly heavy enough to smash it open. Using one rock like a hammer opens the hard nutshell to reveal the goodies inside. It takes a lot of practice.

You need the right sized rock for the right size nut. Too big and the nut becomes dust, too small and the nut remains a nut.

Being chilled

It's a tough job being a chimp. Keeping that clever brain recharged and on top of its game is a good excuse for a little rest and relaxation. Frequently taking time out to chill out is good for a chimp's health. Away from the hustle and bustle of group life, they can settle down to digest a particularly large meal, keep an eye on what the others are up to, think about things and generally rest up.

When chimps are chilling, peace and quiet fall across the group as if they know not to speak to each other. When they do, it's almost whispered, like a chimp library where they will get shushed if they talk any louder. This is generally the calm before the storm.

Chilling-out sessions can end suddenly. A single screech from an annoyed chimp often breaks the silence, kick-starting the whole group into a rousing crescendo of hoots and screams as they react in turn to whatever it was that set them off in the first place. The group goes from being able to hear a pin drop, to a deafening wall of chimp sounds with the volume cranked up to max. And then it's over almost as quickly as it began.

Being almost grown up

When they are between three and a half to four and a half years old, infants are pushed by their mums to do more for themselves. This process is called weaning, and it's tough. After plenty of toddler tantrums and sulking because mum won't always let them get their own way, it's time to grow up.

Once past the weaning stage, young chimps no longer depend on their mum's milk for food, moving on instead to the same varied diet as the rest of the group. They are now known as juveniles, adding new buddies to their growing social network and gradually spending more time with them and less with mum. She might not be as hands-on now, but is always close by. This is part of a parenting plan to encourage juveniles to stand on their own two feet. It's not until they reach the age of six or so that mum's hard work finally pays off and they become independent.

Things really start to ramp up for chimps around the age of seven, with the onset of puberty. Now called adolescents, knowing where you fit in the group matters. Chimp politics helps them find their place and climb the social ladder. Mum will always be there for them though, like waving teenagers off to university knowing that at some point they will be back with their laundry.

Being handy

Chimps have hands that look like over-stretched versions of our own and can use them in a similar way. Long and leathery with black fingernails and gnarly knuckles to walk on, they are not about to be booked for hand modelling jobs any time soon. But they can be daintily dextrous when they need to be and use them in one way or another for pretty much everything – from walking, climbing and gesturing to fighting, using tools and grooming each other down to the finest detail.

Each chimp has a unique fingerprint as well, which is very unusual in the animal kingdom, and what makes chimp hands particularly handy are their opposable thumbs. Just like ours, these can bend and touch other fingers on the same hand, making it easy to grip, hold and manipulate things.

Where chimps have the upper hand on us is in the foot department. Down here they also have opposable big toes, which means they can grip with their feet as well. This makes them perfect for climbing trees and carrying food when both hands are full. Greedy chimps holding lots of extra food with their feet walk with the ungainly style of having a large stone in your shoe.

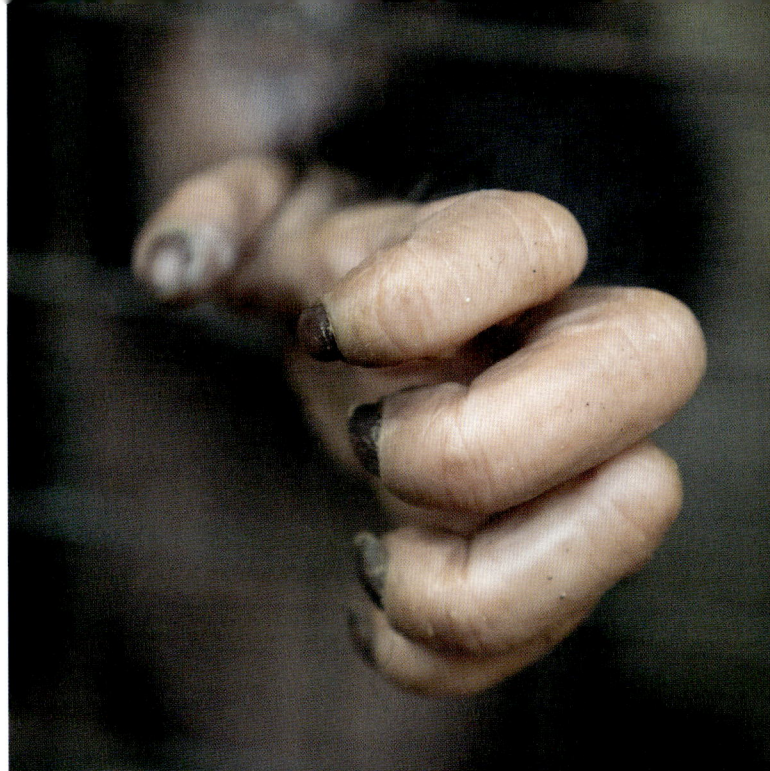

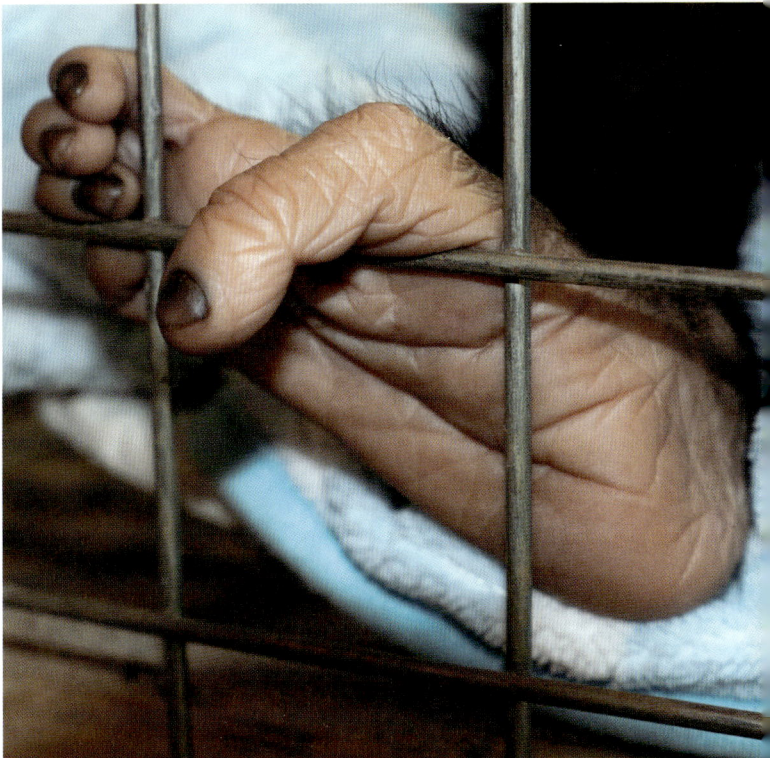

Being able to walk on your knuckles

Chimps walk around on all fours. While many animals that walk like this are called four-legged, chimps are two-legged and two-armed, just like us. But our legs are longer than our arms, so bending down and trying to walk like a chimp is virtually impossible unless you work in a circus.

Chimps have it sussed. Their anatomy is flipped the opposite way round to us, so that instead of having long legs and short arms they have long arms and short legs. When stretched out, their arms are one and a half times as long as a chimp is high, so moving around on all fours is easy, and they always do it on their knuckles. Modified bones in their forearms and wrists lock together when knuckle-walking to support their weight. It also helps that chimps have extra-thick skin on their knuckles for added protection, making their hands an all-weather hiking boot built for rough terrain.

Chimps do sometimes walk upright, or bipedal, but it can look rather awkward; their short legs flanked on either side by long arms hanging heavily to the ground, swaying like pendulums in a grandfather clock. It's no surprise they keep this mode of transport to a minimum.

Being able to climb

Chimps are as at home in the trees as they are on the ground, moving from branch to branch in the blink of an eye. Whether running straight up or bounding on bendy branches, they were born to do it.

Their repertoire includes sliding down tree trunks like a ladder with no steps, arms and legs wide apart, whizzing to the ground in one lock-limbed motion. When faced with thicker, more challenging trunks they adopt the pose of a mountain climber, delicately feeling their way down with one outstretched hand or foot at a time.

It helps that they have very long hands and fingers with short thumbs. Their hands work as a kind of hook when climbing, providing a secure grip and easy release for swinging from tree to tree. Aided by long, strong arms, chimps can swing with the speed and skill of a trapeze artist.

Long arms have other benefits too. Some of the tastiest fruit in the forest grows on branches too thin to support a chimp's weight, but by having long arms, chimps can reach across from sturdier branches to pick the fruit at will. The more they eat, the more seeds they spread in the process, enabling more things to grow. It's the perfect reward for their gardening services.

Being hungry

Chimps are perpetually peckish and snack throughout the day. Breakfast and a late lunch are the main mealtimes though, when the group pack away as much food as they can. But lower-ranked chimps may have to wait their turn. The boss is first in line at any forest buffet and woe-betide anyone who jumps the queue.

They are omnivorous, happy with a mainly plant-based diet topped up with something meatier. The less fussy are known to munch on around 200 different types of food. Their absolute favourite is fruit, but they also eat leaves, seeds, nuts, flowers, eggs and tree bark. Insects like termites and ants are eagerly eaten and packed with protein, and chimps adore honey - often working out clever ways to get to it without being stung.

Meat is also on the menu from time to time, with monkey top of the list. The males work as a team to catch them, moving silently through the forest glancing up at the treetops. When the time is right, they spread out and climb different trees to drive the monkey into an ambush by a single chimp to finish the job. He then gives out portions to his favourite chimps first before sharing with the rest.

Being really grown up

Chimps become adults at about 10 to 13 years old, initially as sub-adults to gently ease them into grown-up life. But from 13 to 15 things start to get very serious indeed.

The boys have a duty to defend their territory from other chimps looking to muscle in, and they do this by working together, forming a chimp army that will fight to protect their homes, friends and families at any cost. The best places to eat in a territory are worth their weight in gold, so chimps defend their favourite restaurants with their lives. While most males tend to stay with the group for life, on rare occasions, the odd one will leave in search of pastures new.

Unlike the boys, when the girls grow up, many are already thinking about leaving home. It's best to start a new life away from related males if they want to have babies of their own, so females tend to pack their bags from the age of 11 to 13 and head off to join another group. Leaving home is not without risk for females looking for romance. Joining new groups often leads to girl gangs beating them up, but making a new friend here and there will eventually pay off and help them to settle in.

Elderly chimps that have been there, seen it and done it pass on their wisdom to the next generation.

Being taken

It is against the law to take chimps from the wild to keep as a pet or anything else. Baby chimps are undoubtedly adorable, very cute and something a family might consider nothing more than a hairy doll for the kids to play with. Some people are willing to pay big money for a living, breathing chimp toy. Other chimps are kept for entertainment at roadside attractions or as something to have your photo taken with on holiday. This new life is no life for a chimp.

But to get a baby chimp out of the forest, other members of the family are often killed first.

Bundled into a dark crate and sold around the world as part of the illegal wildlife trade, some babies are so young and so badly affected that they barely live as long as the flight they are on.

Their new life often sees them beaten and forced to perform, fed the wrong foods, dressed in ridiculous clothes, having teeth pulled out to stop them from biting or locked in a tiny cage too small for them to stand up in. Baby chimps taken from the wild are abused and neglected, sometimes not intentionally, but it is abuse and neglect just the same.

Being abandoned

Baby chimps that are taken from the forest have a short shelf life. They grow up fast – the cuteness factor quickly replaced with a hugely strong individual that is impossible to control.

Growing up without the support and guidance of other chimps, they can become frustrated, unpredictable and destructive. With an innate wild nature bubbling beneath the surface, they often start to lash out. By now everything has grown, including their teeth. Chimps have the kind of teeth a lion would be proud of – their dagger-like canines are powerful weapons that can cut like a knife through butter.

The thought of a chimp with a chip on its shoulder marauding around the house is generally too much for its owners. Inevitably, chimps find themselves no longer wanted, often chained and locked away with no exercise and no stimulation to fuel that wonderful brain. Abandoned and alone, chimps are dealt the double whammy of losing a family twice. Many are subsequently given away, either consigned to a life behind bars or the welcoming arms of a rescue centre where they have the chance to start a new life all over again.

Part 2
Being helped to be a chimp

Being rescued

Sometimes chimps need a helping hand, and it is the babies that find themselves in the most trouble. Fortunately, there are dedicated rescue centres and sanctuaries throughout Africa that do all they can to help. Chimps that were taken from the wild and illegally kept are brought to them to start a new life, learning to be a chimp all over again.

At rescue centres, people take on the role of foster parents. They become round-the-clock chimp carers for those in need of lots of loving attention. The youngest chimps are sometimes so badly distressed that it takes an age for them to come out of their shell.

Forming a strong personal bond with their chimp carer is the best way to help them recover. They will learn everything from them, including what to eat, how to climb safely and how to socialise with their own kind. Eventually, they are encouraged to make friends and form social groups with other rescued chimps.

It is a long road to rehabilitation and one that takes patience and all the time in the world to get right. Rescue centres and sanctuaries are special and vital places. They are chimp champions, devoted and driven to help chimps in any way they can.

Being seen by the vet

When chimps arrive at a rescue centre, they have to be seen by the vet first. Many of the illnesses we get – including colds, measles and hepatitis – can easily spread from people to chimps and from chimps to people, so a full health check is essential to make sure newcomers aren't carrying anything that will make the other chimps or the chimp carers sick too.

It's also the perfect time to assess their overall condition. They are weighed and their heart and lungs are listened to for anything out of the ordinary. Blood tests are also done to make sure everything else is in good working order. The vet can then tell if anything needs treating immediately or simply keeping an eye on.

A spoonful of jollop is often enough to get rid of an annoying sniffle or tickly cough.

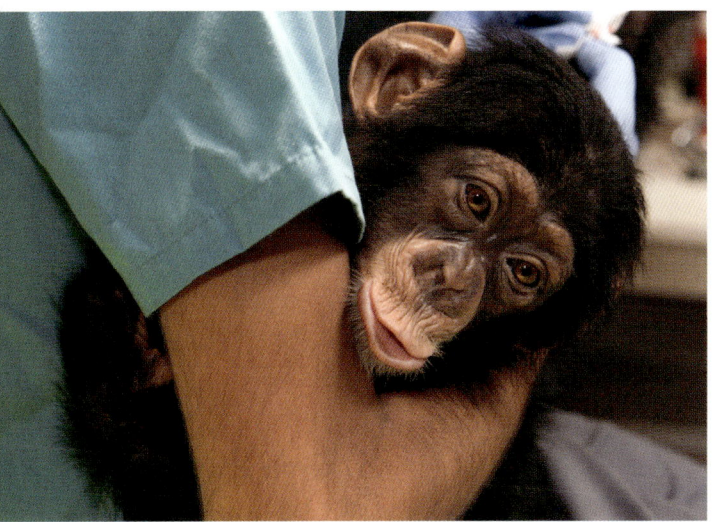

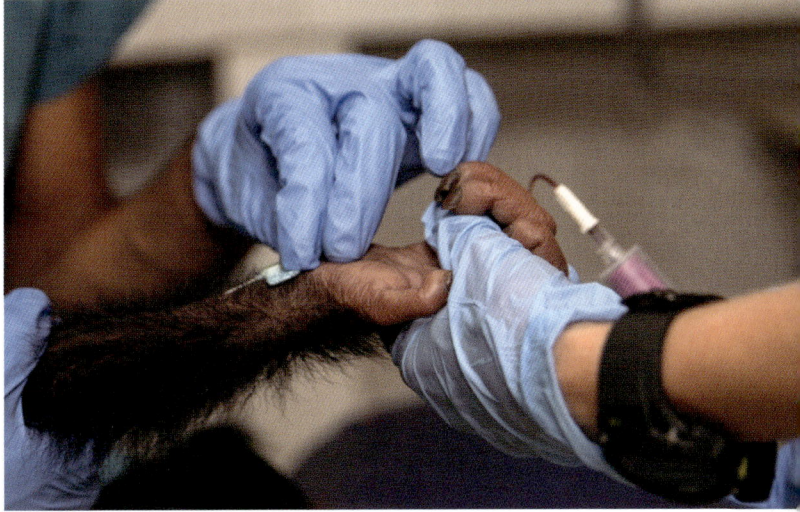

After the initial health checks, newly arrived chimps are kept away from the others for a few weeks just to be sure.

In the wild, chimps are known to self-medicate. The forest is one big medicine cabinet full of prescription plants that they eat to make themselves feel better. Eating leaves from one plant might be just rough enough to push worms out of the gut, while those of another might help with tummy aches.

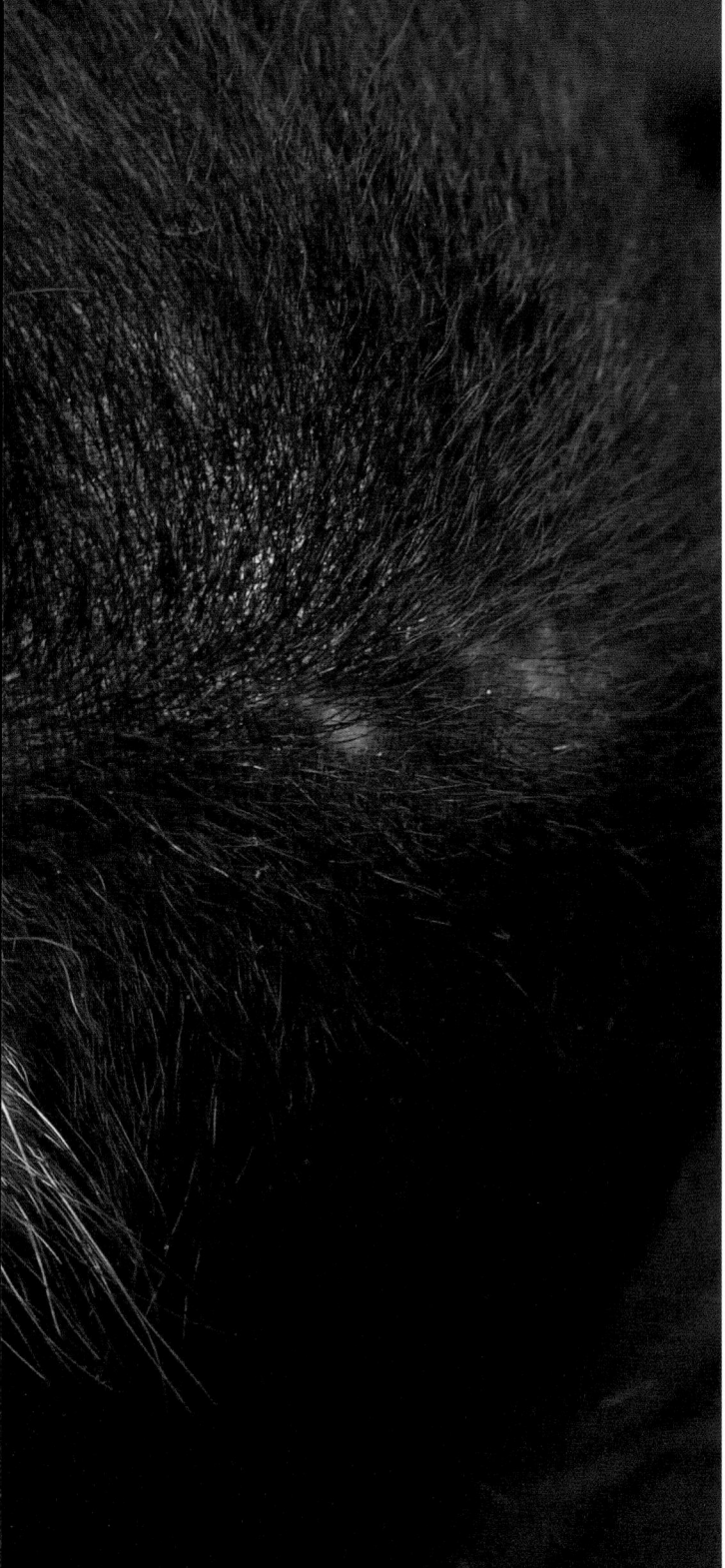

Being sad

For a baby chimp, there is nothing worse than being separated from your mum. They probably saw some terrible things happen to their family when they were taken, and this only adds to the painful memories. Being emotional, with feelings like us, they can quickly go into shock or a deep depression that might last for days or weeks at a time.

Baby chimps usually have a face full of fun, wide-eyed and whirring with wonder at everything around them. This is not so for a sad chimp. Alone and without their family, their eyes lose that sparkle, as if someone switched the lights out. It takes a lot of love from carers at rescue centres to get a sad chimp to switch its lights back on.

Being supported

Rescued chimps need to feel safe as quickly as possible. Some older ones might arrive full of confidence and ready to take on the world, but most, particularly the youngest, are shell-shocked and emotionally distressed by what has happened to them. Devoted chimp carers give them all the support they need.

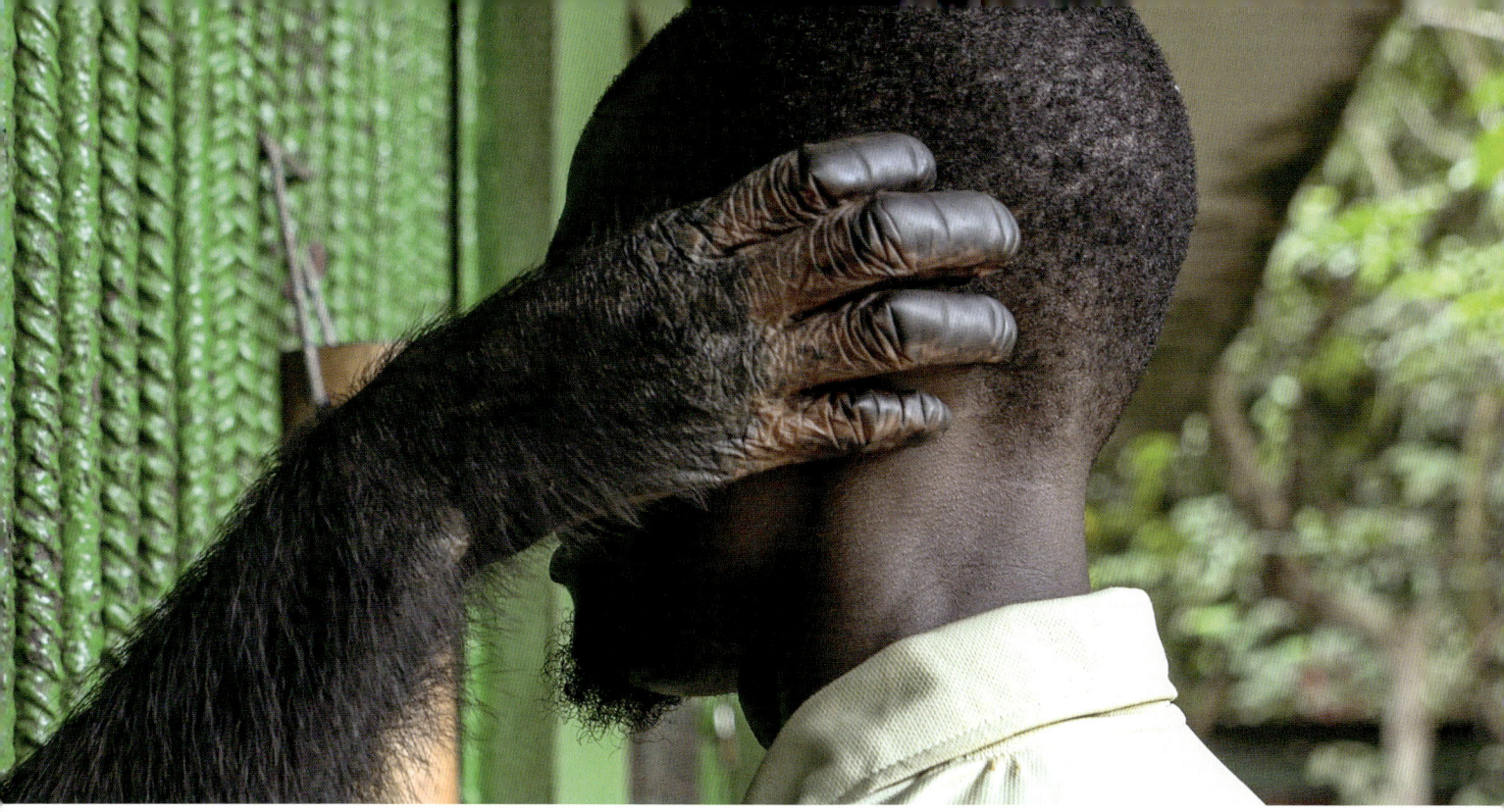

Too big for a cuddle but not too big to say hello to a friend, reaching out to a chimp carer is a touching reminder that chimps are incredibly forgiving.

To help young chimps develop properly, you must get them to trust you first. Forming that invisible bond that says, 'If you think it's ok, I'll give it a go' is so important to their recovery.

Once the chimps start to get their confidence back, they are introduced to other chimps to figure out how to behave in their company and make friends. Gradually, they begin supporting each other more and more until they get most of what they need from their own kind rather than people. At this point chimp carers take a back seat, serving food during play breaks and only stepping in when necessary.

Being bottle-fed

Young chimps rely on milk for the first three and a half to four and a half years of their lives. In the wild, babies carried by mum only have to turn their heads to get a gulp of the good stuff. But for rescued babies, milk is delivered in bottles when their tummies start to rumble.

Picked up, propped up and topped up, the youngest are given milk up to five times a day. It is the perfect fuel and contains everything they need to help little bodies grow into big ones.

Those that need a bit more than others are woken at night for a refill. They don't seem to mind though and are more than capable of guzzling down a bottle no matter how late it is – even when they are still half asleep.

Being cleaned

A chimp mum would groom the grime away in the wild. At rescue centres, grooming babies by hand takes a lot of time so they speed up the process with bathing.

A little accident here and there, particularly in the absence of a nappy, calls for a bowl of water and a flannel. Nappies are put on the youngest to help keep on top of things, but for how long is anyone's guess. Baby chimps have a knack for removing them, often long before they do the job they were put on to do in the first place.

Cleaning really helps the babies relax and they would happily be pampered all day long without a banana break.

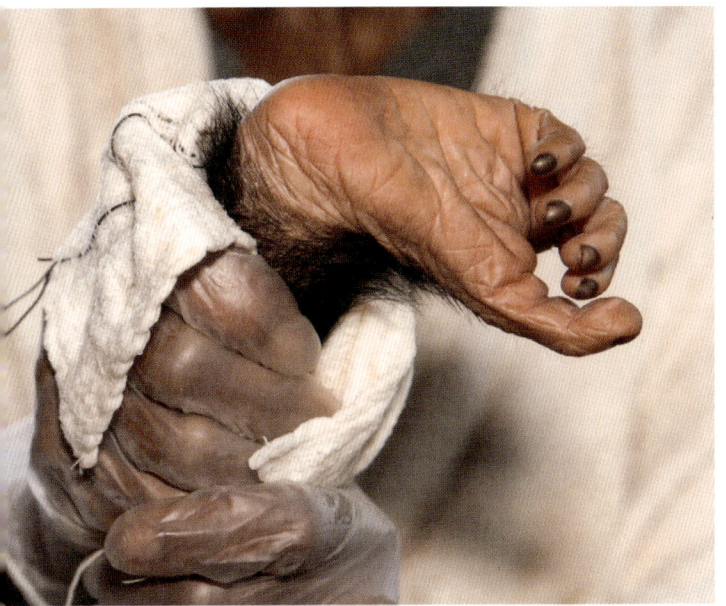

Hot towels are perfect for massaging little legs ready for some climbing exercise.

Wiping away a stubborn bit of breakfast is all part of a top to bottom cleaning service.

Being dependent

Because baby chimps are with their mum pretty much all the time in the wild and depend on them for everything, chimp carers must take on the same role. By caring, cuddling, carrying and supporting them in everything they do, chimp carers form that special connection with baby chimps that is vital for their survival.

At rescue centres, chimps are not the most houseproud of guests and certainly don't keep their bedrooms tidy. Toys are constantly left in the middle of the floor and fruit skins are scattered around like sweet wrappers from a midnight snack. Chimp accommodation is cleaned daily, and things are picked up, put away or replaced.

Chimp carers provide the constant contact babies need to feel secure. They are attached to their carers – quite literally – and hang around watching to make sure everything is cleaned properly. A chimp carer's job is hard enough without a chimp inspector looking over your shoulder all the time.

Being scared

When young chimps are first rescued, they are confused and often very nervous. In the wild, mum cuddles things better, so without her to hug things can get very scary indeed. They need close contact all the time to help them stay calm, but this is difficult when a chimp carer isn't always around. Enter the comfort blanket.

Little chimps love a comfort blanket, either in its traditional material form or a more upmarket woollen jumper version. Both are warm and soft and reassuring to snuggle into until someone returns for the next shift.

Chimps get very attached to their favourite comfort blanket. When not carrying it everywhere, hugging and cuddling the life out of it, they can be found riding it backwards and forwards like a soothing rocking horse.

Being encouraged

Rescued chimps need a little nudge here and there to help them become more independent. There are big trees waiting to be climbed when they learn how to do so properly and loads of healthy foods to eat once they know what they are. Forest school is the best place to learn.

A slice of watermelon is a great pre-workout energy boost. With loads of annoying pips that get in the way, it's a tricky, sticky snack that takes a bit of getting used to.

The forest can be a daunting place at first. It all joins up into one big climbing frame that is as wide as it is tall. But the chimps are encouraged to give it a go, their carers becoming personal trainers gently pushing them to exercise.

Less sporty pupils edge slowly along branches, mindful of the ground below, while the more confident ones go for it with gusto.

Having not worked out the difference between a left-arm grab and a right-arm release, some are destined to drop. For the time being anyway.

Being sleepy

Growing up is hard work for young chimps so they like to have power naps throughout the day. What with learning and climbing and playing and everything, it's no wonder they need to frequently recharge their batteries. A combination of cuddling carer and cosy lap is guaranteed to make even the most hyper chimp nod off.

Milk has a similar effect. Staying awake after a warm bottle is like staying awake to watch television after Sunday lunch. It's not going to happen.

Wild chimps build nests in trees to sleep in. They fold branches and leaves inwards to make a bed comfy enough to snooze the night away. Chimps are fussy when it comes to these sleeping arrangements, making a new bed every evening regardless of the work that went into building the previous night's leafy hammock.

Being happy

When you have the kind of face a chimp does, it's hard to hide your feelings, particularly when you are happy.
A happy chimp can't help but grin like a Cheshire cat, which would make them useless at playing poker.

Rescued chimps find it much easier to settle in if they can start being happy as soon as possible – it is essential to their recovery. Being super intelligent and highly emotional, they need stimulation to stop them from getting bored or depressed.

Chimps react to each other and to what's going on around them, finding fun in the simplest of things and often breaking into the kind of belly laugh normally reserved for a top-of-the-bill comedian.

Being ticklish

One thing guaranteed to get a reaction from a chimp is tickling. Chimps have ticklish spots like we do and pressing these buttons correctly and at just the right time sends their happy meter off the scale. Squirming with a broad, toothy smile while laughing and panting the kind of breaths that could blow up a balloon in double-quick time, a tickled chimp is a hysterically happy chimp.

Being hugged

Who doesn't love a hug now and then? For baby chimps it's not so much now and then, more now and again and again. Hugging is so important to them, particularly the youngest, who are habitually hugged by their mum in the wild.

For orphans at rescue centres, chimp carers are just as huggy and will hug babies for as long as it takes. This is by far the best way to help them relax and feel safe and secure.

Chimps are very touchy-feely throughout their lives and hugging has a special way of making things right. Adult males get terribly wound up at times, but a quick hug from a friend or relative is often enough to calm them down. If a chimp community worked the same way as ours, Doctor Chimp would prescribe hugging for everything. Feeling anxious or stressed? Have a hug. Upset, scared or shocked? Have a hug. In need of emotional support or want to make and keep friends? Have a hug. Injured or poorly? Have a hug. Repeat as required.

The middle of the night can be pretty scary for orphan chimps at a rescue centre. A reassuring hug from a friend is often enough to get back to sleep again.

Being curious

Chimps watch things to learn things. Big brains have a lot of room to store information, so it pays to be constantly curious.

They start learning from a young age, watching everything that the other chimps do and perhaps working out better ways of doing it. Chimps study all the time, even as adults, self-taught through trial and error or by watching headmasterly chimp teachers patiently pass on their many years of knowledge.

Rescued chimps are just as curious. Whether it's staring at the chimp next door or whatever is going on outside, watching the world go by is infinitely interesting.

Being curious is all part of an ape apprenticeship where a sore thumb from bad tool use, a smack from a grumpy chimp, or being sick from eating something bad is all part of the process. In the long run, being curious is the surest way to being successful.

Being independent

As rescued chimps grow in confidence, they start doing more and more things for themselves, spending a lot of time exploring and working things out.

No longer fully reliant on their carers, they climb high in the forest on their own. By now little gangs of chimps have become buddies and are forming new social groups, jumping and playing in the trees together like seasoned pros.

Tasty takeaways are all around and working out which ones are good to eat is all part of growing up. Leaves and shoots are the equivalent of forest fast food – quick and easy to grab on the move. Some fruits are a simple peel-and-push-down-the-throat job, while other snacks often need a bit of problem-solving to get into.

When eating, if a chimp pulls a face like it wants to sneeze, it probably doesn't like the taste of something. If, on the other hand, a chimp looks deep in thought as if filling in a crossword puzzle, it definitely likes the taste.

Being endangered

It is thought there were once as many as a million chimps living in Africa, but the world has changed considerably since then. The amount of people living on the planet is eye-wateringly high and this has had a huge impact on chimps. So much so that they are now endangered.

While the human population grows, so too does development around chimp habitat. As a result, forests have been cut down for their wood and to clear space to grow food, raise livestock and mine minerals. Environmental factors and their effect on habitats and ecosystems are also a concern. So too is hunting.

Animals that are hunted from the forest to eat – called bushmeat – become more and more accessible the more and more trees are cut down. Sadly, this includes chimps. Live animals also sell for a good price, so for a hunter, catching a baby chimp at the same time is a bonus.

Chimps need all the help they can get. Luckily people can rethink, forests can regrow, and chimps can recover – if given the time and space to do so.

About the author

Adrian Cale is a wildlife filmmaker, writer and naturalist.

He has filmed, directed, produced and written wildlife TV documentaries for many different television channels including BBC, Animal Planet and National Geographic, and has contributed work for various conservation organisations.

He has raised butterflies to film their life cycle, dressed up as a giant panda to blend in, been pooped on by camera-shy monkeys, charged by elephants, slept with a deadly puff adder and shower-danced with scorpions.

Adrian has also appeared on our screens as a wildlife TV presenter and has narrated numerous documentaries.